GALACTIC FRIENDS

The CE-5 Coloring Book
(close encounters of the 5th kind)

Designed, Illustrated, and Written by Rae Dove

I0482550

Spotting a UFO

A beach sighting!

Together as one!

Galactic friends

After a trip outdoors, connecting with nature, and offering Star Visitors peace and friendship.

Pleiades, a star cluster in the constellation, Taurus.

Thousands of stars in the clear desert sky

A friendly sighting

Helping to keep Earth clean

Earth is a beautiful place!

Star Visitor caring for Earth

Starships and UFOs

Seven Starships

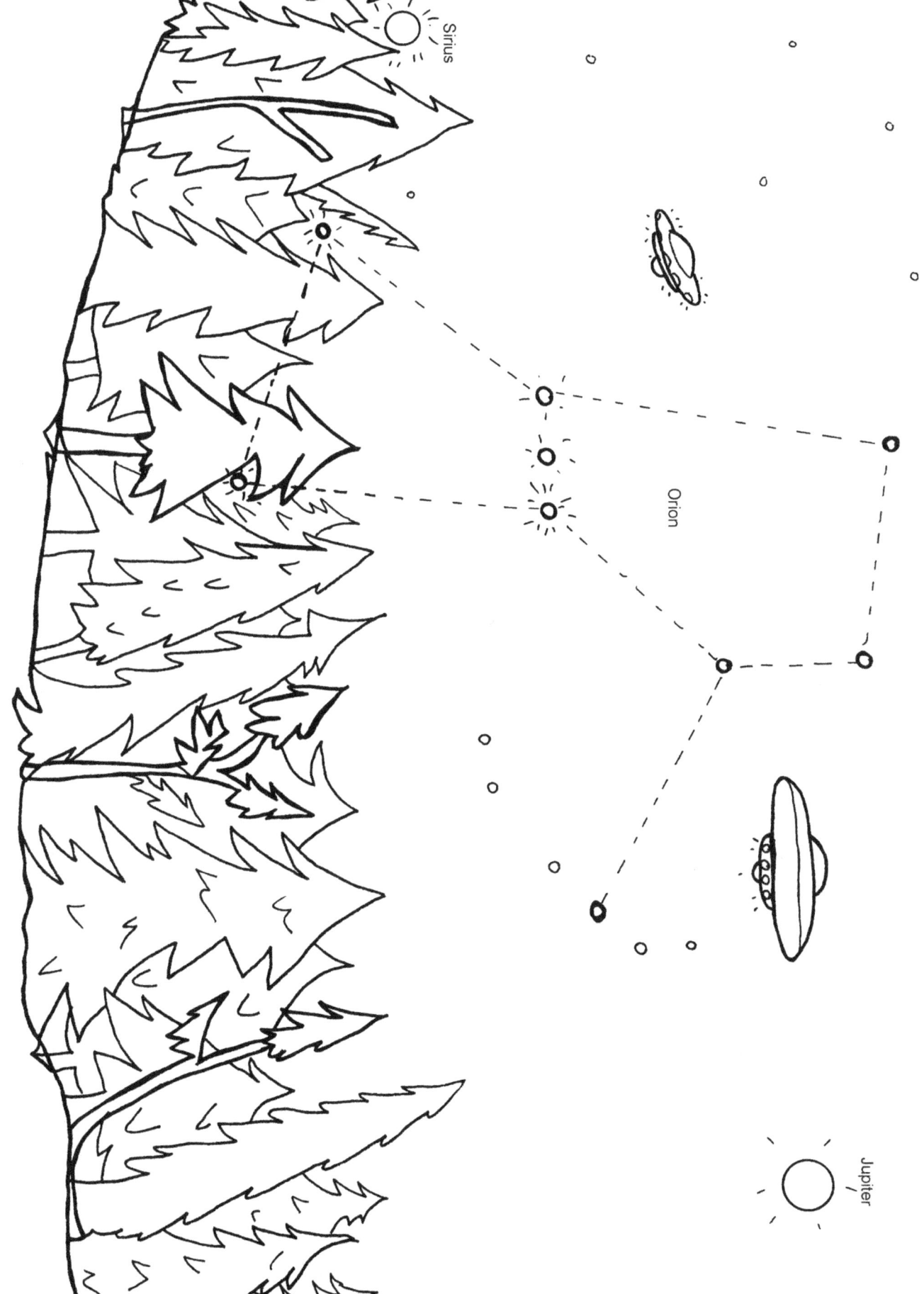

The star system, Sirius, planet Jupiter, and the star constellation, Orion.

Sirius

Orion

Jupiter

A friendly Star Visitor

Peacefully Meditating

Peacefully inviting Star Visitors to say hello

Star Visitors

Peace starts at home.

Space Doodle